CW01475960

The Almond of my Mind

the
Almond
of my
Mind

The Poetry of Neuroscience and Love

Anna Ehrenman

Cover illustration by Gabrielle Havens

Human anatomy and cell biology review by Abigail Lawton, PhD

Book design by Julieta Raimonda Silva, Ariel Vandebrooke, and Nichole Delaura

To the *WANDERING* ones,

as they S E A R C H for their place

to *BLOSSOM.*

Table of Contents

Acknowledgments

Many loving thanks to my sister and brothers —
Esther, David, and Michael — who fostered
creativity in my life through our own little "Bronte
Writing Club;" my parents for providing an
education that cultivated my mind and influenced
my work; my publisher Ariel for seeing a poet in
me; my friends — the Marauders and Gazelles — for
their inspirational, brilliant artistry and cherished
friendship; and last but not least,

my flower garden of a human.

Introduction

I have worked in gardens since my earliest memory. In elementary school, while other children played with toys, I spent my weekends tending plants and my weeknights reading botany books about the healing, toxic, and even deadly uses of the plants I was growing. Botany is the study of how best to shelter and cultivate life—both in the plants themselves and in the people who may benefit from them. It is a world of knowledge and cures that I adore.

By 7th grade, I had discovered another such world: I added neuroscience to my reading list. I had always been intrigued by the formation of thought, movement, sensation, and memories — in this new field I found the keys for all four. In the study of neuroscience, I discovered that through remembering and understanding, the partially known can lose its horror.

I spent time in Australia during my childhood and adolescence. Australians are forced to reckon with an often forbidding environment. During the 2019 wildfires, I witnessed ash falling from the sky onto my outstretched hands and resting snowflake-like on my hair. The experience will never leave my memory. Nature, at once, can be captivating and cruel. So can love.

In the summer of 2021, neuroscience brought me a new — and much more unexpected — form of healing. As I floundered through a period of unusual emotional stress, struggling to make sense of and "re-set" the broken pieces of hope, fear, happiness, and regret within me (as a surgeon might re-set a bone), I suddenly found metaphors staring back at me from every page of my neuroscience textbooks.

A favourite poet of mine, T. S. Eliot, said that the world ends not with a bang but a whimper; but what if it also falls apart one small snip at a time — growing ever shorter like the telomeres inside our DNA? Shakespeare said that the beloved is like a rose, just as sweet by any other name, but what if the lover also is the almond of our minds, the amygdala that colours our memories with emotion? Piece by piece, the botany of the brain seems to consume me, driving me to commit a little thievery of the fourth dimension by clacking out these poems on my laptop.

Neuroscience makes me as excited as meeting a loved one every time I think or read of it. I hope you too will find love in these pages — maybe even a cure.

Notes on the Science

---

* LIGHT RECEPTORS · Noun [laɪt ri ˈsɛptər]

There are two types of light receptors:

1. Cones are the ones for colour vision and high light situations.
2. Rods show only black and white and are for low light situations.

See " Bursting colour in the midst of black & white"

The centre of my mind

Refers to the Hippocampus which is fundamental in memory formation.

Insufficient rate

The Foxglove flower contains a chemical called digitalis that is used for slowing one's heart rate.

Almond of my mind

The amygdala, whose name comes from the Latin word for 'almond,' is a small brain region responsible for fear responses and associations, and thus is integral to the emotional experience of memory.

A Flower Garden of a Human

You are the *iris* of my eye,

The greeting petals

Bursting c o l o u r in

The midst of black&white,[*]

The wrapping 'round

The CENTRE of my mind,[*]

Specked all through by

Inherent vice,

No blemishes of dirt

Smeared on you by

The world about.

You are the *rosehip* of my throat,

The barbed branches

Wrapping round it

Like an inverted collar,

The s c e n t of roasted

Rose hips touches

My lips, a scented kiss

EPHEMERAL

Upon my senses,

But e v e r l a s t i n g in my mind.

A sip of rose hip tea

Would incapacitate me,

A virgin brew,

Yet still INTOXICATING

.

You are the *foxglove*

Of my little hands & heart,

The strapping stalk

Catching my eye,

Leeching out words

Before w h i s p e r i n g away

In a flash of black&white ,

For the *iris* leaves

My life in shades of grey.

Yet now, the speckled petals

Fall from my fingers,

But the lingering

Effects upon my heart

Are evident in its

I n s u f f i c i e n t R a t e.*

You are the *almond* of my mind,[*]

The KERNEL of emotion,

Encapsulator, captor

Of my sharpest memories.

I could fill an ENCYCLOPAEDIA

With how you fully

Encompass *flowers*

With your properties,

Your effect upon me,

But I'll settle for these few,

My flower garden of a human.

IRIS

n, pl. irises or irides [ˈaɪrɪˌdiːz; ˈɪrɪ-´]

1. (Anatomy) The contractile membrane perforated by the pupil, and forming the colored portion of the eye.

2. (Plants) A genus of plants having showy flowers and bulbous or tuberous roots, of which the flower-de-luce (fleur-de-lis), orris, and other species of flag are examples. See Illust. of Flower-de-luce.

3. (Class. Myth.) The goddess of the rainbow, and swift-footed messenger of the gods.

4. *A symbol of sight, perception.*

Notes on the Science

---

* NEURON · Noun [ˈnʊr ɑːn]

  The cells of the nervous system that receive sensory input and then transmit the information throughout the body.

* AXON · Noun [ˈak-ˌsän]

  A portion of a neuron, often long and threadlike in shape. When the axon of one neuron and the dendrite of an adjacent neuron join, a synapse is formed.

* SOMA · Noun [ˈsō-mə]

  The cell body of a neuron, containing the nucleus.

Vile Accomplices

Refers to dendrites, branching areas of neurons which receive the electrical or chemical messages from axons.

This Fragile Fabric of Memory

My thievery

                                        of their plasticity

              would crystallise

        These axon's* *use to one and one alone,*

Your visage

                        —the neural* fabric coruscating

              At me with your likeness.

        *The axons' vile accomplices in tow—**

Those threads which

ever trace to you,

I'd pluck them out

to wind around my distaff.

Their incessant blinking captured in a woven image,

Soma[*] hung along it,

pearl-like in their innocence.

Notes on the Science

---

* DEPERSONALIZATION · Noun [diː,pɜːsənəlʌɪˈzeɪʃən]

A disorder that makes people feel as if they are observing themselves from outside their body. They have a sense that things around them aren't real. As if they were living in a dream.

The Warped Speed of Perception

The mirror caught my eye

a little while,

                              A little while too long,

                                        it seems,

For my accepted concept of me

and that reflection

                              became irreconcilable.[*]

The consciousness

                              And sense of self

                                        divorced

By nothing

but a sheet of plastic

                                        catching

Light

                        and shunning it,

photons

                                    immigrating

To one's eyes

to strengthen

horror's sense.

The mind discerns

the nature twain,

And shivers itself

in fractured

sentience.

The image

and the self

are not the same.

Notes on the Science

Steeds

Refers to the neurons.

Compass

The magnetic material in homing pigeons' beaks which allows them to sense the Earth's magnetic field.

Part One: Of Pigeons and Potters

My focus falters, my mind shifts,

The steeds* bearing thoughts stumble,

Undergrowth springs up

Upon the paths, obscuring them.

I'd flattered myself

To be a guiding beacon,

But now I find

My own feet

Wandering with aim

But no direction.

I realise my compass*

Sets itself with points obscure

And threatening,

when I follow them

Blindly, for I trust

Something other than

The maker of the North.

The needle shifts,

The points of places

Set at the beginning

Of the Earth shift into place,

My vision clears,

As do the paths beneath my feet.

The earth is green but guiding

When it is not relied,

but trod, upon.

Notes on the Science

---

* OCULAR DOMINANCE · Noun [ˈɑːkjəl ər ˈdɑːmɪnənts]

  The propensity of one eye to be favoured over the other very slightly.

Relentless Night

The twain of us were cast together

Into darkness, resting nervous in

A condition foreign to our function.

But we, in huddled harmony,

Continued commerce in charges,

Hoping for relief from darkness,

But understanding that we must fire

In order to survive, to keep reviving

Those functions for which we formed.

Had one in light to light ascended,

We would have lost our companionship —

The one in darkness lost its sight,

Remained,

but an abandoned path

On which

no new impulse

would traverse.

But our shared trial maintained our roads,

Despite the gloom incessant we cleared

The brush and bandits, filled the divots,

Drew anew the lanes to keep the nerves

Upon their predestined paths.

Our destination received us both,

Encouraging us as we spurred each other on

still,

The humming of the other well-used senses

Sometimes stirring envy in our seemingly

Senseless plodding, hoping for the stimulus

To return and bring the news

To our occipital centre.

And then

                                        —behold!

Together we captured dawn,

The first dawn witnessed

since the ageless midnight fell,

And down the maintained roads,

The rosy blood pooled

On our masterpiece,

Beaming sweet encouragement on us.

It was not for naught!

We cried out in delight.

Our process fine-tuned

In all the lack of light

Our companionship

Had borne us through

relentless night.

Notes on the Science

---

* BASILAR MEMBRANE · Noun [ˈbæz ɪlər ˈmembreɪn]

A structure that extends from the base to the apex of the cochlea, dividing it into two chambers. It vibrates in response to incoming sound waves.

Base Distraction

I wake wrung out, my palms are chafed

From REM-free cycles staging out —

The waking worries endlessly.

The wringing winding 'round my life,

Caught in the careless grip

Of bandit thoughts and wayward connotating.

A ringing fills my ears,

As if the basilars* created oceans

Of their own volition

To counteract a void of noise

Within this blossom'ng envelope of silence,

A silence ringing like a siren

As pent-up,

                    uncontainable,

                        unmentioned thoughts

            Beat wave-like against clear walls.

Notes on the Science

Nothingness

The ability of the brain to receive sensory information and not create a conscious experience of it occurs during both sleep and wakefulness, but primarily in sleep.

The Mind Demands
More Certainty

Frankly, I love my honesty,

But should honesty require of me

Vulnerability,

I flee immediately to ambiguity, Whose vaporous arms conceal The ill-defended points. *A billowing emptiness pours forth from me,* Consuming sounds and sights and smells and words and tastes and touch, But never gnawing on the stimuli,

Absorbing them into nothingness.

Notes on the Science

* TELOMERES · Noun [ˈtel ə mɪrz]

A protective feature that prevents the degradation of terminal regions of chromosomal DNA. When the telomeres are too short, the cell will no longer form offspring. While telomere shortening is associated with aging, it may not be causative.

* MOTOR CORTEX · Noun [ˈmoʊʧ ər ˈkɔːrt eks]

It is located at the top of the brain, and is responsible for generating the signals that execute voluntary movements.

* NOCICEPTORS · Noun [ˈnoʊsɪsɛptəs]

Receptors that detect the sensation of pain.

Each cycle

A reference to the endogenous clock which keeps the roughly 24-hour Circadian rhythm.

"Winding from the motor cortex to nociceptors of my feet"

A reference to the phrase "from head to toes."

The Hands of Time are Shrinking

There's a sing-song warning from the clock within,

Its shifting face conveying with each cycle: *

> *'You're running out of time, what is your life?*
>
> *What have you done in life,*
>
> *now that you're running out of time?'*

I hear it echoed in the clipping ends of telomeres * —

Each time I heal snips them closer to their end of use.

Relentless rhythms

tapping at their stations

tapping at the glass,

rives and changes

thoughts' courses

to match their own,

Winding from the motor cortex* to nociceptors* of my feet,

Daily winding down and 'round until the muscles seize

And cease their function, the fibres instead pulsing with

The questions trickling in from doubt above:

> *'You're running out of time, what is your life?*
>
> *What have you done in life,*
>
> *now that you're running out of time?'*

Confronted with the question of

the culmination of our lives

the mind turns interminately

to contemplate the termination,

So goes the way of life in contradiction of itself

When Socrates or his imitators in the self arise.

There's that tapping at the glass

which, with time and desperation,

Blossoms into fists

which strike

with wanting might,

A bird forever striking

on the glass

        to reach the warmth and life it sees within.

There's no success,

of course,

the warmth and life are lies

To lure the hopeless bird,

a false promise of some respite

From the clock

and telomeres

in cruel and constant concert.

For, life lies not within or without, but above,

The level gaze finds only gloom,

the upward gaze can pierce it through

To find the beaming face, a sweeter voice

Calls or whispers:

'I guide you in my time, this is your life;

Bird-like, live your life,

I guard and guide your time beneath my wings.'

And though the telomeres trim

and clock creak on,

The voice remains

and soars in northward song.

ROSEHIP

n. pl. rosehips [ˈɹoʊzˌhɪp]

1. (Plants) The fruit of various species of the genus *Rosa*, especially those found in the wild; the fruit generally appears red but can range from a light orange to a purplish black.

2. (Medical) A supplement derived from the fruit of a rose bush and considered beneficial to the health of the heart.

3. (Food) A fragrant tea.

4. *Themes of new love, desire, passion, or emotion.*

Notes on the Science

---

* SULCI · Noun, pl. [ˈsəl-ˌkī] (sing: sulcus)

The valleys or indentations between the folds of tissue on the surface of the cerebrum.

Foaming Future

The crystallised blue sky above

       reminds me once again

Of new horizons yet unreached,

      and stirs the swirling press

Of future's weight

    upon my back,

       the former steps

Left crumbling

      into fractured forms

    of memories

Poorly

preserved,

a honey for my mind

To melt within

Its mouth

with ruminating

As the night,

the promise of another day,

descends on me.

THE SHATTERING SKY BEHIND ME

fills my paths and pockets,

Overflowing thoughts and sulci[*]

with their bitter shards

And hidden sweetness,

for the moment passed in greenness

is once more

snatched

From out my life,

and out of reach,

and ne'er regained.

THE FUTURE CHUNDERS US,

its foamy peaks ever in pursuit,

'til the rip comes in

so soft,

So suddenly —

and tears the breath and future

from our limbs.

Notes on the Science

---

* SULCI · Noun, pl. [ˈsəl-ˌkī] (sing: sulcus)

  The valleys or indentations between the folds of tissue on the surface of the cerebrum.

* OCCIPITAL LOBES · Noun, pl. [ɑːkˈsɪp.ɪ.t̪əl loʊbs]

  A region of the brain responsible for some of the processing of visual information from the optic nerves.

Sulcus·

The thought of you swims effortless

Forever in my mind,

So much so that they would find,

(Should some sculptor pierce my skull)

T h e l e t t e r s o f y o u r n a m e

Spilling from the sulci cavernous,

Recesses gushing forth coagulates

Which formed the name—fermenting,

Swirling, a tea of unfulfillment

Filling up the cup made from this skull.

And should they drink of it,

Then you would swell as well

Within their thoughts,

        Until the sight of you would

the occipital lobes[*] from

                over-welling sulci

                    SPLIT

Notes on the Science

---

* ANTERO-RETROGRADE AMNESIA
  Noun ['antərō're-trə-ˌgrād æm 'niːʒ ə]

  An uncommon form of memory loss in which the sufferer
  only has short-term memory, meaning they cannot form
  and retain any memories for longer than several seconds.

Antero-retrograde Amnesia*

My mind swims while eyes

Are dry and bright,

I cannot recall

'Fore falling thrall

To your wondrous soul,

My life, its all,

Is swallowed now.

If you thought

aloud,

I would know whether

My words linger

In your mind, belied

By glazing eyes

Fixed on empty walls

After duty, me — and others —

Rend us two apart for spells.

I cannot think

Or recall other times,

And I believe

That with each BLINK,

This is all we are, or ever were.

Notes on the Science

———————————————

.

Pull No Pavlovs*

I stand amidst the rain, how warm,

How odd, in near mid-autumn.

I hear your call, and run to meet

The one who calls, on feet not fleet,

But ever eager.

You are stimulus,

unconditioned,

I never had to learn to set

My mind on other things

When your cry would sing,

Like the cardinal who calls

His other half, as red leaves fall

My swift uplifting

Rush upon these borrowed wings,

An unconditioned

response

for I need no reward at all

My heart comes swiftly

When you call

Notes on the Science

* AMBER · Noun [ˈæm.bɚ]

Norse mythology tells of Freyja who, when her husband was away, would cry tears of gold and amber. For centuries, yellow amber has been used by natural healers to improve memory. Amber is considered a gem, though it is not a stone. Rather, it is a fossilized resin from the sap of ancient pines.

Amber Darkness

I am an insect trapped in amber.*
I scream, but the world can't hear to answer
Forever encased in a beautiful haze
I watch - through a warped blur -
Them live out their days
Frozen in resin, hard as stone
I gasp, and I suffocate alone.

Notes on the Science

—————————————————

* SYNAPSE · Noun [ˈsaɪ.næps]

A junction between the axon of one neuron and the dendrite of an adjacent neuron. The synapse utilizes chemical or electrical conduction. Electrical synapses allow for rapid signal transduction, while chemical synapses allow for slower signal transduction. Electrical synapses use a type of cellular connection called "gap junctions."

The Nature of Building Trust

At this juncture,

I've concluded

That lasting connections cannot be

gap junctions,[*]

For they fire endlessly until worn out,

No inhibition,

no mitigation of their strength,

                              No consideration of the other,

Only immediate knowledge and babbling

— 'Til there remains no future.

Rather, we must wait   for time   and use

To accumulate acceptance and perception

To guide our chemical connection[*]

To uptake, the synapse strengthens

Until our subtlety gives way
Until we cast off scents of amber
Until the almond blooms in fullness

Opening to trust,

Opening to day.

Notes on the Science

IOLITE · Noun [ˈīəˌlīt]

Also known as "The Viking's compass," Iolite is a gemstone that Old Norse navigators used to reduce cloud-glare in order to find the sun's position.

NEUROTRANSMITTER · Noun [ˌnʊr.oʊ.trænsˈmɪt̬.ɚ]

The neurotransmitters VASOPRESSIN and OXYTOCIN are crucial for empathy and social bonding. Some people are genetically predisposed to have fewer receptors for oxytocin, and as a result, have difficulty showing kindness to and forming strong bonds with others.

"Vastly pressing" a play on vasopressin.

GLUME · Noun [ˈglüm]

A leaf-like structure that encloses flowers. Specifically, either of two empty bracts at the base of the spikelet in grasses.

A play on the word "gloom."

POPPY · Noun [ˈpɑː.pi]

A herbaceous plant grown for its colorful flowers. The poppy is a symbol of eternal sleep since one of its species (Papaver Somniferum) is the source of the drug opium.

MICROGLIA · Noun, pl. [mī-ˈkräg-lē-əl]

Microglia are a type of neuroglia located throughout the brain and spinal cord. They act as the first and main form of active immune defense in the central nervous system.

NIGHTSHADE · Noun [ˈnīt-ˌshād]

Due to high levels of alkaloids, the Solanaceae family or nightshades include many toxic and even deadly plants.

You Lack Components
Of Compassion

YOUR BRUSQUE FRANKNESS
not yet brutish

—punctured the amber glume[*]
Which held me fast,

And I delighted in your cutting company.

You fashioned iolite[*] lenses for me
—to pierce the haze of misery's resin,

but I spied their crystals creeping o'er
my already amber-encrusted skin,

VASTLY PRESSING[*] ON MY THROBBING TEMPLES,

Iolite LAID SIEGE upon my breath,
Sealing my lips
F r o m s w e e t o x y g e n.
I saw you laugh.

You laughed to see me suffer so

Beneath our duet
                    of cruelty

And I knew
—*despite the amber's whispers*—
That I must stem the source of agony.

THE AMBER WOULD HOLD AND SUFFOCATE ME
For eternity should I allow it

And you
                    — in pointed pursuit of entertainment
                    — buried me in a tomb of your containment

Deigned to bypass

AND SEIZE CONTROL OF ME.

Against my carotids
          poised the iolite,

Hawk-like in its predation,
Poppy-like* in its sedation,
          piercing in my mind
          by creeping up my spine

I identified the Brutus
though it pained me
And in the sticking place,
Seized your arm in false embrace,
Drew you in so swiftly

THE FORCE OF OUR COLLISION

shattered   your   iolite

And with it, the idolization of your presence—
Serration — once sweet— soured in its curdling.

You staggered back, your face a sneer,
no sympathy,
no regret,

and soon I c a u g h t a g l i m p s e

of you prying through another amber-smothered victim,
And I knew
I'd been but another in your queue.

Bonds between humanity mean nothing

To you, for you feel nothing

Or very little

I swept aside the iolite, and sighed,

But lo!

The amber

shivered, c r a c k e d, and wailed,

And I pried it from me
As an old man sheds his skin.

I crack sweet breath between my teeth,
PAIN — *it withered!* — in my breath-fed smile.

I rise from cleaning your encapsulation,

Leaving you to harvest

choked blossoms of your cruelty,

You burrower of brains,
A mole of microglial[*] nature,

YOU NIGHTSHADE[*] OF A HUMAN BEING.

FOXGLOVE

n. pl. foxgloves [ˈfɒksˌɡlʌv]

1. (Plants) Any plant of the genus *Digitalis*. The common English foxglove (*Digitalis purpurea*) is a handsome perennial or biennial plant which can cause an extreme allergic reaction when handled without gloves.

2. (Medical) The leaves of a *Digitalis* plant, which are used in the medical field to create sedatives.

3. (Poison) Any plant of the *Digitalis* genus containing *digoxin*, which is toxic to the heart.

4. *Toxicity, betrayal, and death.*

Notes on the Science

---

* ASTROCYTES · Noun pl. [ˈæstrəʊˌsaɪt]

Cells that provide structure, nourishment, and filtration of blood for the nervous system. The man who named them included the root word 'astro' so future researchers could easily identify the star-shaped cells.

Cerebral Conquest

Moving with the rhythm

of your heart to guard

Your brain and spine,

so nourishing,

I am your cerebrospinal fluid.

I bypass astrocytes

With bribes

of lingering affection—

This much is true, I swear.

So that your brain may bathe

And ponder in me,

And I may seep into your

Every sulcus.

Notes on the Science

---

* BRAIN STEM · Noun ['breɪnstɛm]

The brainstem consists of the medulla oblongata, pons, and midbrain. It connects the spinal cord with the brain. The main function is to control the vital body functions. The three parts of the brainstem regulate the involuntary functions.

1) MIDBRAIN · Noun ['mɪd breɪn]

Processes information from the eyes and the ears. It is also responsible for motor coordination response, pain, and behaviour towards fear and anxiety.

2) PONS · Noun pl. [pɑːnz]

Links the medulla oblongata and the thalamus, the pons send information about sensation, motor fuctions, taste, hearing, and eye movement.

3) MEDULLA OBLANGATA · Noun
[meˈdʌləˌɑːb lɔːŋˈgɑːt̬ə]

Continuous to the spinal cord, the medulla oblongata sends signals between the brain and the rest of the body. It is also responsible for speech and taste.

*The one who controls your
brainstem controls you.*

Brainstem Blues

You have a string

in my medulla [*]—

      I'm convinced by now

The sight of you, it crowds

Out instinct—breath is robbed

      Cruel, clinging master is this thing.

      I gasp; You tug upon the string,

Still my heart.

My blood flow

Stops

Again,

a thread becomes yours in my pons [*]

I am blind

to the other cardinals that call

A blur — they fly past my periphery

They cannot catch my eyes

for I am caught

Already in the lies of your sly beauty.

You turn;

You tug two strings at once,

And I can't help but leap and run

Towards your voice—

You know you own me

You hold my brainstem;*

You control me.

You have my medulla,

midbrain,* and pons,

my love

You are my skullcap,

my passionflower,

and my foxglove

Should I be wise now and sip carefully?

If I should drink to excess–

                              death.

Do you understand?

Regardless, you hold my strings within your hand

I dare to flirt with fate;

I dare to let you linger,

I cannot be reticent;

Reach out to wrap you 'round my finger,

Let us have a taste

Of each other's medicine.

Notes on the Science

_____

"*The clock and wires within your head*"

Clock refers to the telomeres within each cell which are believed to have a link to the ageing process. Wires refers to the neurons.

Irony Along The Flower's Veins

The blossom peaks in vibrancy,

The call of death rings higher pitched,

Sometimes in the midst of life

Until it grinds the green to grit

The petals buckle 'neath the weight,

And form the yawning pit

In which the live one flits

In vain escape from tug and toss

From life and death

—in their violent new attempts

To hold the lasting sway upon them.

The clock face, like a blossom past its bloom,

Grows withering and ripples in itself

—*too soon*

Mov' endlessly about their track

Despair, it glistens

on the blade of hands

The end is known, you know, I know,

The clock and wires within your head[*]

Keep close their track of fading days

As telomeres trim our lives of grace.

The hands, they moulder.

The face, the blossom past its time—

      They fold in on themselves.

      The clock by its master is dismantled,

      The blossoms return to dust from which

      They sprang.

      The bees, with their honeyed flight,

      now fed or dead...

Regardless of the time, our sun will set,

Our morning flowers fade

No time, No Joy, can keep sweet death at bay.

And into that ageless midnight—

We, petal by petal,

decay

Notes on the Science

---

* DORSAL ROOT · Noun [ˈdɔːr.səl ruːt]

    Dorsal nerve ganglion neurons originate from the dorsal root of the spinal nerves and carry sensory messages to the central nervous system, such as touch and pain.

Grass Shouldn't Be Green in December

Grass shouldn't be green in December,

Flowers shouldn't bloom so soon before the snow

But there beyond the glass they grew and lingered

Headless in devotion, heady in their growth

Envious in shade,

but of what I can't consider.

Darkness fell and, silent,

Swept the flowers from my sight

At midnight

came a splitting

'tween my brainstem

and the cord.

The dorsal roots*

of spinal nerves

sprung the vertebrae

upright

The roses wept and withered

In the frost that fell tonight

The callosum body shuddered

Betwixt the regions twain

The flickering between them

Began instantly to fade

I SAT IMMOVABLE

CORTICAL CONVERSATION CEASED,

The green had cried out warnings

Of bizarre, unusual fates

For there,

Beneath my motionless chin,

There shivered

Still a spear

Firm lodged

Within my chest

And through the wall

The fragile petals, frozen fall

And I

Claw at this winter, shivering

Twixt my medulla, midbrain, pons–

The rosehip's gone.

Notes on the Science

---

* SYNAPSE · Noun [ˈsaɪ.næps]

A junction between the axon of one neuron and the dendrite of an adjacent neuron. The synapse utilizes chemical or electrical conduction. Electrical synapses allow for rapid signal transduction, while chemical synapses allow for slower signal transduction. Electrical synapses use a type of cellular connection called "gap junctions."

Little Girl

Against my aching chest I press, The golden head all *filled*

With a bureaucracy, Of phobic thoughts and ways,

Of searing pain you fear to say—

      s i l e n c e

No longer, The nerves between our minds melded by

Gap junctions,* Your pleading tone leaking sore

Through my synapses* from yours—

      h o w t o b e s u r e

DEAR CHILD,

I am your metamorphosis,

Not quite the final form

Your present problems will resolve

fall

AWAY

beneath the blades

and blinding flames

You will learn: Of growth and friendship,

God's saving insight: *to be understood*,

REASON *and* EMOTION–

I know, the last two are your dearest friend & most loathed foe

Golden head all honey-soaked, Forsake Plato

And his foolish chariot, For he misunderstood how horses

In companionship should run–

t o g e t h e r

Oh foxglove heart, allow, embrace emotion,

Not pure reason only, For without one the other is turning

Aimlessly in self-concentric circles–

    c o m e s u n d o n e

                    Dear child,

                   Let pain not be your friend,

                   Nor Death your lover from

                   Your age

                   For he is still foremost in my heart.

        S L E E P

now, your sweet and weary head, lies heavier

you will find your light within this midnight

Notes on the Science

———————————————————

"Sculptor of the clay"

*And the LORD God formed man
of the dust of the ground,
and breathed into his nostrils
the breath of life;
and man became a living soul.*

*And the LORD God
planted a garden
eastward in Eden;
and there he put the man
whom he had formed.*

*Genesis
The King James Bible*

Of Pigeons and Potters
Part Two: Reprieve

Why praise the vessel

When the former of the clay

And sculptor of the clay* from naught

Envelops us, unseen?

Embrace the former of the earth,

Who moulded you from it

And then endowed you with

A breath,

In its strength — of life and soul

so terrible

so tender

Notes on the Science

---

* LIGHT · Noun [laɪt]

  The brightness that allows things to be seen. Human sight occurs when visual receptors capture this light. It also initiates the process of photosynthesis.

* SHADOW · Noun [ˈʃæd.oʊ]

  A reverse projection of the object blocking the light. A silhouette.

Shadows are a symbol of soul, darkness, and transcendence. They also represent secrecy and the unknown.

The Sleep-Deprived Reprise of Visions

The yellow p a r a l l e l s stretched out beyond

The reach of lights* and cones

The *bristles* of the pines withering from green

Into black as the man-

Made monster lit the trees, brighter than moonlight.

My hand rested LIGHT upon the wheel

The faux leather's lack of texture melting

into a slender ***shadow**** shoulder

OF YOURS.

As I recoiled, a SPEAR appeared through my windshield—

Not shattering it, but with the force

OF SIX HUNDRED FIFTY-FOUR NEWTONS

It *pierced* my forehead through–

The shaft s h i v e r i n g in my skull.

YET

—when I looked up—

I saw and felt your lips against my skin,

Your hair falling almond dark upon my golden,

My shattered forehead, prisoner to that bitter kiss,

Your gentle press upon the puncture-point

Unmaking, rejoining, healing, breaking,

Brains and bone, my skin

Undulating indecisively, shifting

Between wanting healing and hoping

FOR THE HURT TO L I N G E R.

The lines before me *c u r v e d*—

I know not how—

but my hands guided

'Round the bend

myself and YOU

The yellow p a r a l l e l s stretched out beyond

The delight and terror

The longed-for spear & your visage had shattered,

Mere *shadows* across my lap and shoulders, leaving

Nothing but an ache behind my eyes and in

The *centre* of my mind.

ALMOND

n, pl. almonds [ˈɑːmənd]

1. (Etymology) From the Latin "amygdala" - the same word used to describe the almond-shaped mass inside each cerebral hemisphere. It is mainly through the amygdala that humans experience fear.

2. (Plants) The fruit of the almond tree.

3. (Medicine) Bitter almonds are a poisonous volatile oil.

4. *Memory and possession.*

Notes on the Science

---

* DOPAMINE · Noun [ˈdəʊpəmiːn]

A neurotransmitter and a chemical messenger that allows communication between the body, the brain, and the nerve cells in the brain. The hypothalamus releases this hormone into the bloodstream which creates the sensations of motivation and pleasure.

High and low levels of dopamine are associated with diseases such as anxiety, insomnia, addiction, and aggression.

Sate Heaven's Notions

I seek your blessed company

TO SATE SWEET HEAVEN'S NOTIONS,

The seeding spark, it gleams—

Your voice,

your face,

SETS MOTION

Through   *the bitter*   *system,*

Of undulating,   *grinning smug emotions.* *

You inspire
a bleeding
sweetness,

Cruel and sour honey-oil of almond

Now   *at last*   *succumbs*   *to rest*

In long-awaited, long-foretold dilution.

SATED HEAVEN'S NOTIONS,

Leave me not, you're but a swan
Of newborn confidence.

No hiss
escapes the grove
now honey-grown,

Angel-bees have kissed and blessed
Thoughts' sources input, out-gone,

SOME HEAVEN'S NOTIONS SEEDED,

BLOSSOMED & DISPERSED.

Notes on the Science

---

* PREFRONTAL CORTEX • Noun [ˌpriː.frʌn.t̬əl ˈkɔːr.teks]

Dopamine in the prefrontal cortex modulates cognitive control, thereby influencing attention, impulse inhibition, and prospective memory. It is implicated in controlling aspects of language, speech, and the capacity to differentiate conflicting thoughts.

The Shape of Time,
It Shifts About You

The shape of time,

                                        it shifts

                about you,

Rippling lower, swifter, in your standing wake.

The hushing rush of honeyed time encircling

Me

            entwines

                                        itself

                        into

            my

                                            hair,

the
c r y s t a l l i z i n g
threads becoming...

...*i n d i s t i n c t*

As, in returning kind with kind,

the t h r e a d s   *to* t r a c e   *y o u r* s u l c i

n o w I *w i n d,*

P

I

E

R

C

E

*your brainstem
& prefrontal cortex*[*]

So to bind your thoughts to mine

Notes on the Science

_____

Begonia

This perennial flower
symbolizes kindness, gratitude, or caution.

Baby Blues

Tender sensitivity, innocence and trust.

Roses

Pink roses symbolize appreciation, gratitude and
admiration. Red roses symbolize love and courage.

Violets

In Ancient Greece, these purple-blue flowers were seen as a
symbol of loyalty, modesty and purity.

Asters

These daisy-like flowers were used by the Greeks on the
altars of their Gods' as a symbol of love.

Babies Breath

Represents innocent love, new beginnings,
and happiness.

A Limbic Bind

I've found no man forgets my name nor face,

               Though boyhood's passed,

YOUR MEM'RY'S NOT.

We *stumble*

       on

        in *reminiscence,*

p r e s s e d   v i o l e t s ,   c l u t c h e d   b e t w e e n   t h e   p a g e

'Til you betray

                 The lingering
                    knot.
               It
            somehow
          seems
       my
     threads
   are
 laced.

w i t h i n   t h e   p a t t e r n   o n   y o u r   l o o m,

m y   m e m ' r y ' s   t r a c e d

o u t   b y   y o u r   h i p p o c a m p u s,

c r y s t a l l i z e d   b y   y o u r   c o l l e c t i o n

Of recollections of insight—linked

With-memories-of-opened mind--

the-revelation's-light.

How odd to be the pivot,

To be the hinge of one's experience,

Perception,

self

-deception,

AND REASONS FOR DEFENSE.

Forget me not,
and I will bind

                       The baby blues
                       and breath

Into my braids
—and mind

                       You carry
                     out the roses,

           I appreciate your thoughts

                     BUT pray you find

                     a WOMAN who

Can let YOU fill her own, and pluck the proffered

Queen Anne's Lace, placing in your open hand

    A cluster of her asters and begonias

Then, may I find peace in clutching these ***aged*** flowers[*]

You once SACRIFICED to me.

Notes on the Science

---

* HIPPOCAMPUS · Noun [ˌhɪpə(ʊ)ˈkampəs]

From the Latin "hippocampus" - denoting a sea horse.
From the Greek "hippos" and "kampos" - "horse" and "sea
monster." The hippocampus within the brain forms long
term memories of events, facts, and spaces in one's life.

*In a tragic but crucial case for scientists to discover the
hippocampus' role, Patient H.M. suffered from anterograde
amnesia after his hippocampus was removed. He never
formed new memories again.*

Essential Seahorse

A mistake uncurdled is a mistake crouching

In anticipation of meeting me once more–

So, slumbering s e a h o r s e,* curled within

The circling walls, curdle recollections–

For no combattance in the self denies it second chances.

To this h o r s e alone I entrust

the dredging up of well-scraped churnings–

Yearnings, sweet and souring in the smatterings

Of *you* remaining; I fear forgetting–

Stumblings into Lancelot-like combat, almond memories.

Fragily, I protect the s e a h o r s e churning,

Swiften all his curdling and over-turnings–

And pour out a drink still brimming

With steam and bitter-frothing lemon–

Lest I betray myself to you *again.*

Notes on the Science

---

* BROCA'S REGION · Noun [ˈbroʊk əz ˈriːdʒ ən]

The motor speech area that regulates the breathing patterns while speaking. Its name comes from the French physician, anatomist, and anthropologist Paul Broca, who studied language-impaired patients and discovered this language area. A damaged Broca's Region results in the inability to speak.

"Wheedling almond"

Certain conditions can cause hyperactivity of the amygdala, such as anxiety disorders.

Sour Almond, Honey

The daily rhythm of your mind must be exhausting,
All the worries, quibbling, and erroneous assumptions

That all others are purposefully malicious against you.
Your thoughts are wound about the wheedling almond,*

Wormwood-whispers poisoning your perceptions.
The almond's oil, so foul in festering, stains your pupils,

Trickles down through your nostrils,
seeps into the roots of speech and teeth.

Broca's region* fell against the almond's onslaught months
ago—Another of its devastating conquests.

OH, TO SEE YOU CAPTIVATED BY A KERNEL OF EMOTION,
THE ALMOND ESSENCE SHARPENING YOUR MIND
TO RECOGNIZE ALL BUT GOOD INTENT TOWARDS YOU.

If I could but pry it from that earth, unearth yourself
Long lost in inner torrents—but alas!

Without that almond, you would lose much more
Than torrid preoccupations.

Please

Spur and source your mind and heart to swell and cleanse
you from within!

I CAN DO NO MORE BUT WIPE
SOUR HONEY TRICKLING DOWN YOUR CHEEKS
AS YOU REBUKE ME FOR MY TROUBLE.

Notes on the Science

---

* MEMORY · Noun [ˈmem.ɚ.i]

  The ability to learn, store, retain, and retrieve information. There are three categories: sensory which is subconscious, short-term, and long-term.

* SHORT TERM MEMORY · Noun [ʃɔːrt.tɝːm ˈmem.ɚ.i]

  These memories usually last up to 30 seconds and are short-term storage ephemeral synaptic connections capturing a moment unless the memory is saved in long-term memory.

* LONG TERM MEMORY · Noun [ˈlɑːŋtɝːm ˈmem.ɚ.i]

  These memories vary in how long they are kept, from several minutes to the rest of one's life.

Echoic memory, or short-term memory, can be preserved as long-term memory with enough attention and repetition.

Gratitude

Should it be I haven't mentioned near enough

Times, in passing or in conscientious effort,

How glad you make me simply with existing

In as close proximity as we can muster:

MY FLOWER GARDEN OF A HUMAN

Your voice *blooms* with turns of phrase,

The dear rising, sacrificial *scent* of your flabbergasted tone
Undone by what I've said

Your echo growing, *budding* till it plays
In LTM[*] (WHEN YOU ARE MILES AWAY)

FOR DEATH WOULD BE SWEETER
THAN MY SYNAPSES' GRIP SLIPPING
ON THE SNIPPETS OF YOUR VOICE
I CLASP WITHIN THE FOLDS
OF MY TEMPORAL LOBES

The simplicity of our interactions brings   a p e a c e   quite
foreign upon me and my little self, this tapestry we've spun
— in themes medieval & modern Tolkein, in sunshine and
in shadows without sun, in buzzing bees & in honey made
up of foxglove, in the way YOU say my name
                    and make me run

Should it be I haven't made you wake and feel

A gratitude similar to mine, that we still share

Each other's space in time, echoing, till distance

Shrinks in strange, swift aging: Listen–

I AM CONTENT

to have your voice herein